Immortal Tomorrow

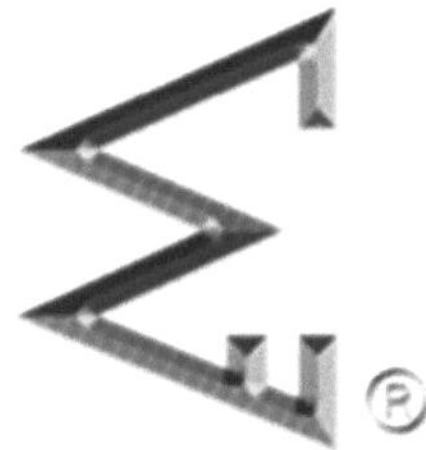

Edward L. Hannon

Content

Edward L. Hannon©

Foreword

Acknowledgement

Philosychology

Existential Concepts

Dark Humor

Foreword

Although, the title of this book is premised upon the reality of tomorrow, the feasibility of its vision starts now. For immortality have two states of being: cyclic and non-cyclic. Cyclic immortality comprises a point of reference that consists of birth, transition, rebirth; but non-cyclic immortality is a vibratory incorruptible frequency, which transcends all space and relative time.

Acknowledgement

I would like to give special thanks to "SOURCE CONSCIOUNESS," which is "ALL THERE EVER WAS, IS AND EVER WILL BE," for the profound intelligence, knowledge and wisdom to create such an awesome work. I would also like to give special thanks to those who contributed to Microsoft Clip Art, which helped offer visual interpretations to my literary concepts. Lastly, I would like to thank you the reader for your time, interest and attention.

Philosychology

Philosychology (noun) is a study of conscious behavior created by PhTCB (philosopher and teacher of conscious behavior) Edward L. Hannon. This science is a synthesis of empirical psychology along with his philosophies to methodize a practical or practicable solution to resolve the dilemmas, conflicts, or queries that mankind perpetuates upon itself.

Chaplain Edward Lewis Hannon D.D. (Doctor of Divinity)

Linear thinking, within the age of the Kali Yuga, has had its saga of limiting potential; but, it shall continue to find asylum in those who are diametrically opposed to advance their way of thinking.

Freedom must prevail; lest a nation's health is reduced to become an invalid of feckless potential.

Some are apt to assume that diplomacy is a gullible sense of displaying kindness; without prudently considering that it may be a last-ditch effort to maintain some semblance of civility.

Stop giving those the status of importance, who have nothing going for them, but try to insult others as though they do.

Once you understand and prudently live by the golden rule, you do not become consumed with seeking revenge; for you know that, by proxy, whatever ill-intentions that someone may direct towards you, typically pales in comparison to the personal agony that they inflict upon themselves.

The best tangible description that I could give for heaven or hell can be utterly defined by one word: "Truth." For however truth is perceived or accepted, determines either the well-deserved state of psychological freedom; or the self-defeating hell of mental turmoil.

In order to dotingly love someone that you do not yet trust, you must first be willing to sensibly gift the practical regulation of implementing some sound sense of necessary structure.

Existential Concepts

Existential Concept: The Linear Paradox for Relative Nonlinear-Frequency Time-Travel

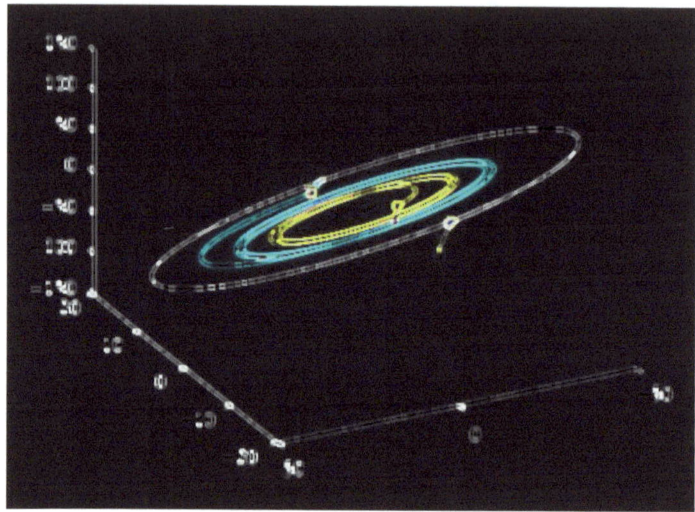

Consider the inner nonlinear circle, along with its three unique points of reference, as the quantum super-predisposition of the past, present and future. Also consider the outer-projected measuring features, as linear points of space/time frequency orientations. Thus, within the Grand "INTELLIGENT DESIGN" of Existence-Itself, time travel, in a preliminary sense, can be feasibly understood as a calculable trajectory of infinitely absolute sequential information; coupled with its further projections of subsequent potential. For, it has been said, "What we do in life echoes in eternity."

Existential Concept: Numerical Chaos and Order

1   2

              +   = 5   "Isfet"

3   4

5   6

              +   = 13 = 4 "Ma'at"

7   8

9 = The Existential Balance of Existence

Existential Concept: Quantum Entanglement Explained

Quantum entanglement can be defined as a kinetic unit or system; which may be dynamically charged by how its polarities, within a given frequency, either correspondingly cooperate, reflect, deflect or repel further acculturations of kinetic potentialities.

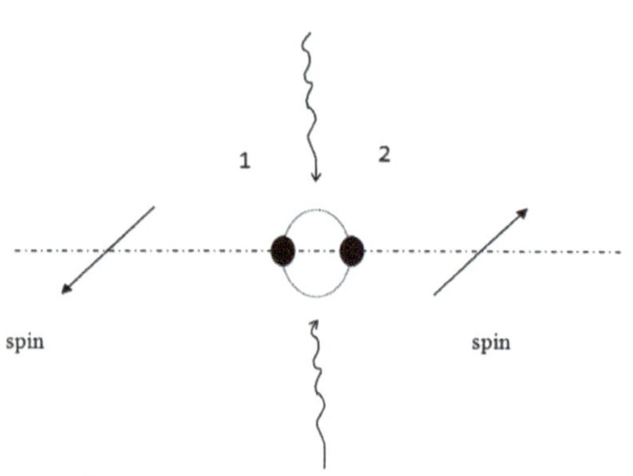

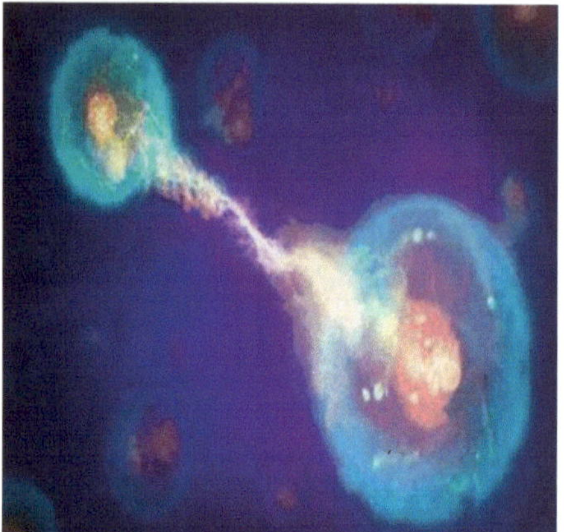

Fig. 1: Production mode intertwined couples destruction.

Esoteric Concept: The Afterlife within Life

Once you awaken from the dispositional sleep of 180 degrees, within the 720 degree crypt of corporeal perception, then entering any subsequent vertical door of also 720 degrees becomes a potential portal for further psychological advancement.

Esoteric Concept: The Dance of the Existential Yugas

Within the illuminated circle, three quadrants of intense radiant light are dimmed by the fourth fractional prospect of an obscure quarter of apportioned shade.

Existential Concept: Dark Matter Explained

Dark matter can be defined as a coagulative force, which links and then harnesses kinetic particle charges; by dynamically grouping their corresponding and receptive properties of conductive potential.

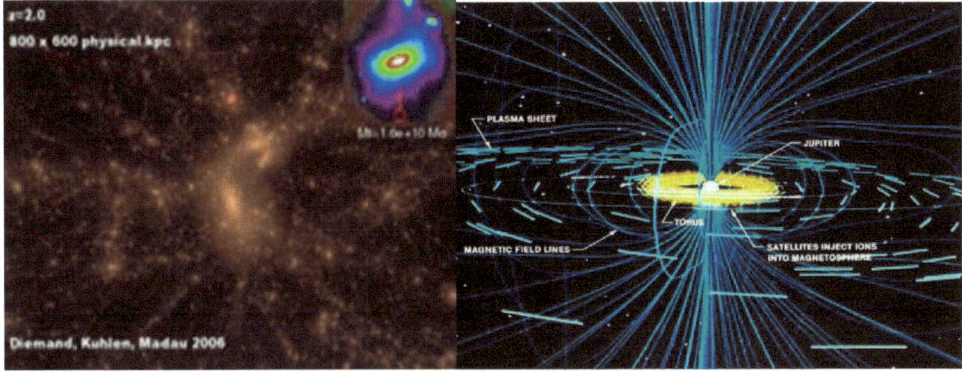

Existential Concept: Immortal Resonance

5.28.9.10 Gamma Mega-Hertz is an immortal frequency of space-time incorruptibility; which can casually pervade all kinetic cultures of energy resonance.

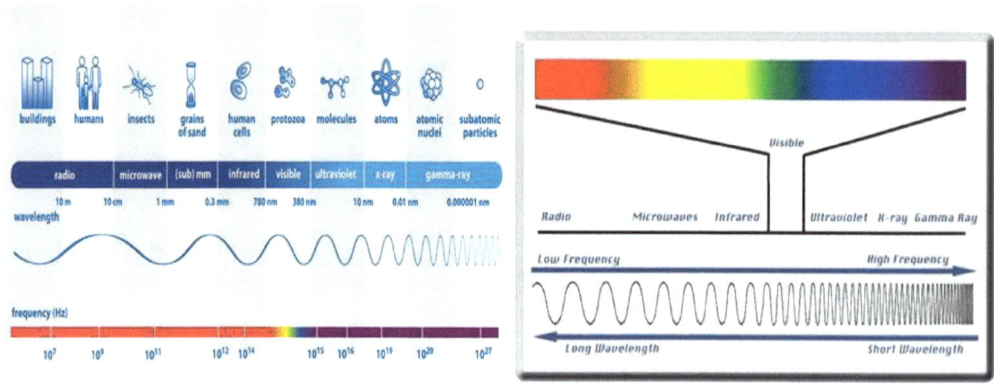

Existential Concept: Initiating Kinetic Variation

As it relates to INTELLIGENT DESIGN, catalytic disruption can be defined as a calculable destabilization of a statically neutral energy-field; by diametrically positioning force-dynamics, in order to facilitate nascent kinetic behavior.

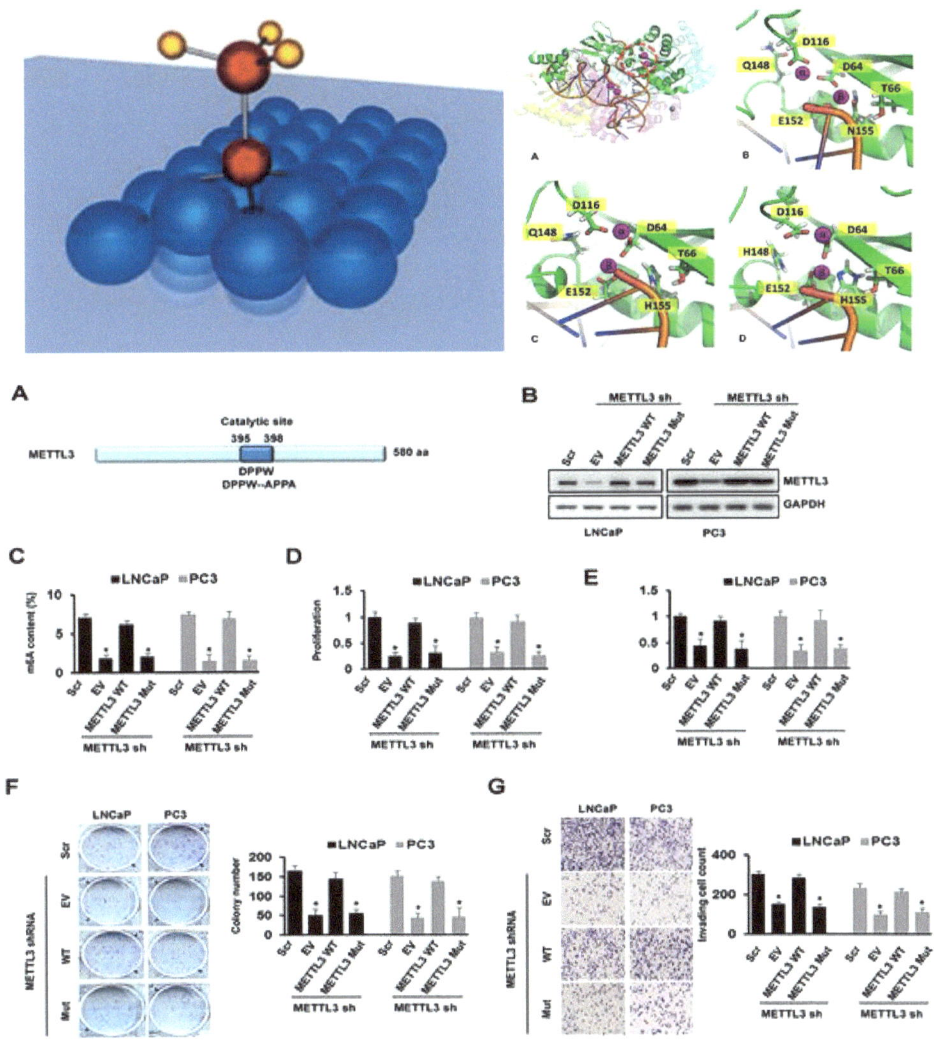

Existential Concept: Transcending Light Speed

Impulse celerity can be defined as a hyper-rate of velocity, facilitated by the ease of how electromagnetic patterns, waves or signals travel; using the kinetic path of least resistance.

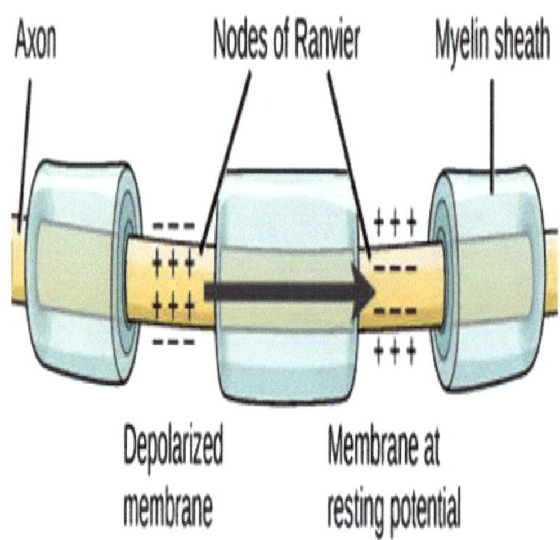

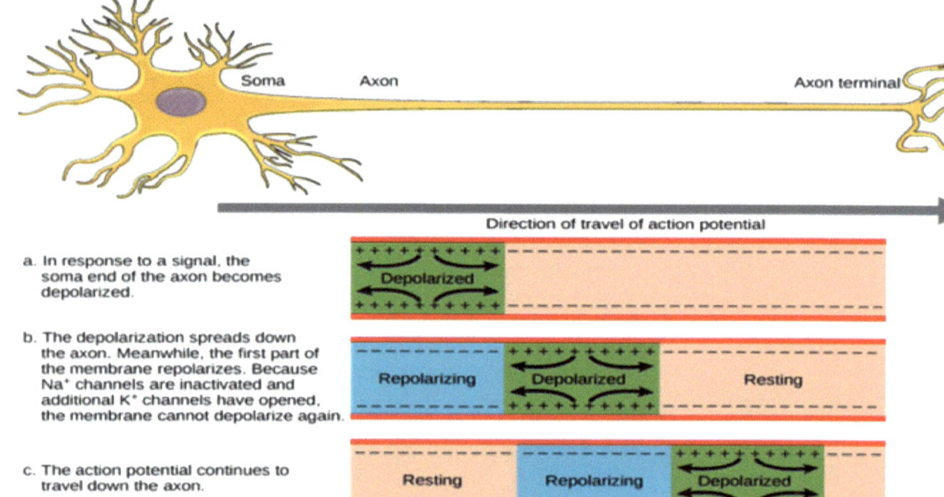

a. In response to a signal, the soma end of the axon becomes depolarized.

b. The depolarization spreads down the axon. Meanwhile, the first part of the membrane repolarizes. Because Na⁺ channels are inactivated and additional K⁺ channels have opened, the membrane cannot depolarize again.

c. The action potential continues to travel down the axon.

Existential Concept: The Adiabatic Microcosmic Universe of the Third Dimension

The nature of the third dimension is a supremely isolated state of existential development; because each dimension must have its unique quality of individual expression. Therefore, extraterrestrial interference from other dimensional planes would be moot or unjust considering that existential natural law allows for the psychological agency of unencumbered freewill to be exercised. But, if there is some form of contact by other dimensional civilizations, then it would typically be by proxy. Now, if this concept sounds utterly absurd, please consider the microcosmic reality of intra third-dimensional relations and interactions between third-world and developing nations, in contrast, to more technological advanced nations.

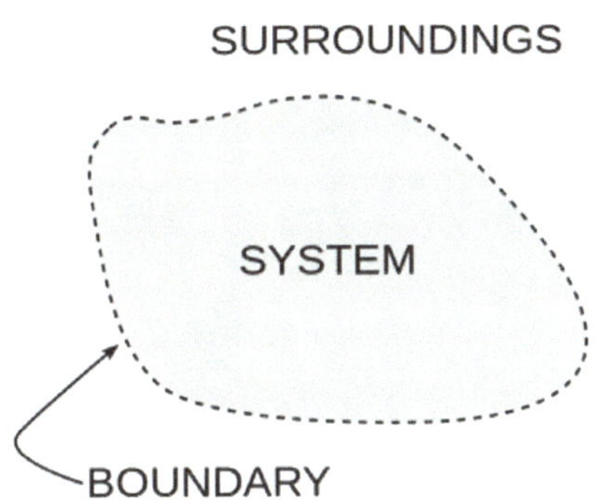

Existential Concept: Confirming Infinity by the Rations of Measure

Some religious schools or esoteric thought say that the earth was created in six days and man was created in one. Rational science has posited that the universe is about 13.8 billion years old, earth is about 4.543 billion years old and man has been on the earth for approximately 230,000 years. Cryptically, I ask, who is man, to dare, to make such an idealized claim, other than a DIVINELY INFINITE ESSENCE; which is numerically mapping an existential estate of reference, within that which is but a mere portion of ITS immensely incalculable QUALITY?

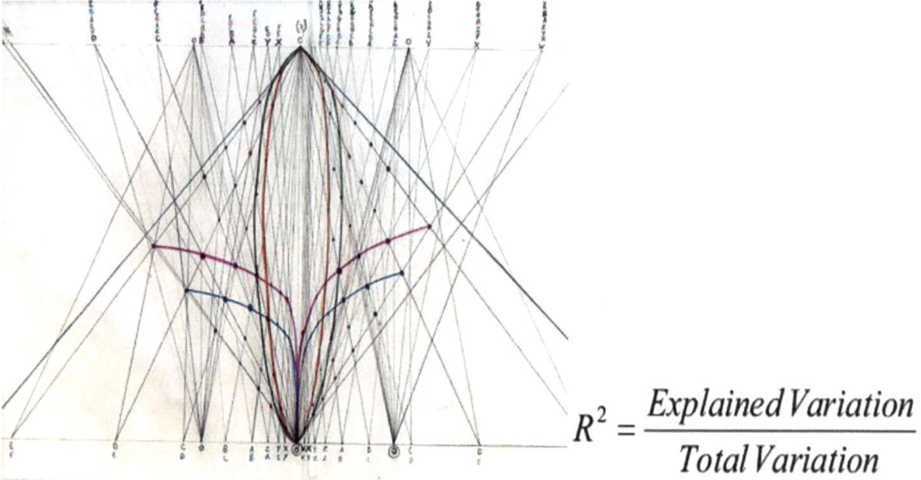

$$R^2 = \frac{Explained\,Variation}{Total\,Variation}$$

Existential Concept: Existence Outside of Conceptual Bias

As it relates to INTELLIGENT DESIGN, beyond space-time orientation, houses initially, a logistical reservoir of indiscriminate kinetic behavior or culturally indistinct dynamic potential.

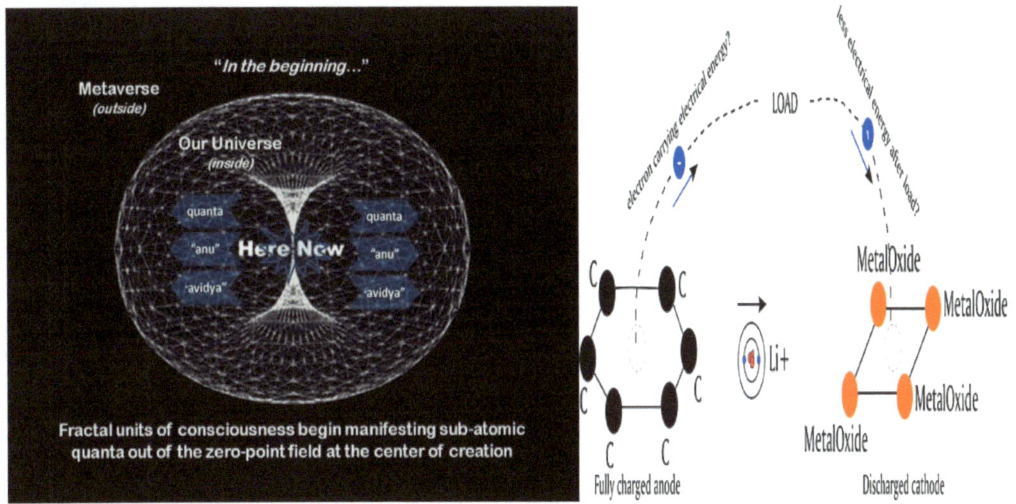

Existential Concept: The Psychological Androgyny of Balance

As it relates to INTELLIGENT DESIGN, once consciousness reaches a gestalt of veritable consistency, then it can emerge as a teleological agent of amaranthine capacity.

 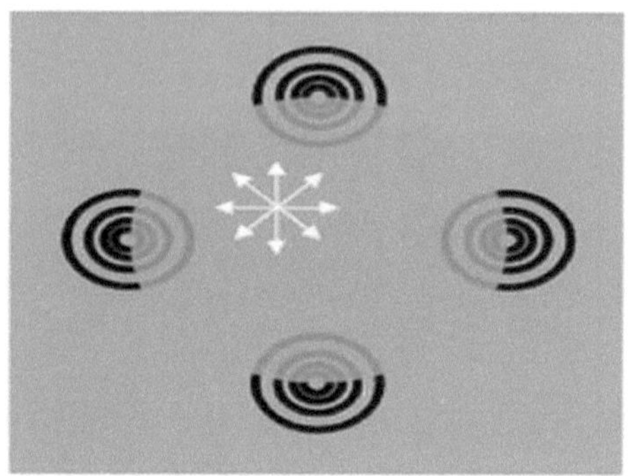

Complementary Properties of Boundaries and Surfaces

Boundary Completion	**Surface Filling-in**
Inward	Outward
Oriented	Unoriented
Insensitive to direction-of-contrast	Sensitive to direction-of-contrast

Existential Concept: Immortal Tomorrow

Density fluctuation, within the ne plus ultra of amaranthine capacity, is enabled by kinetically corresponding with environmental tendencies; in order to either biologically sync or become displaced from the cultural norms and dynamics of energy facilitation.

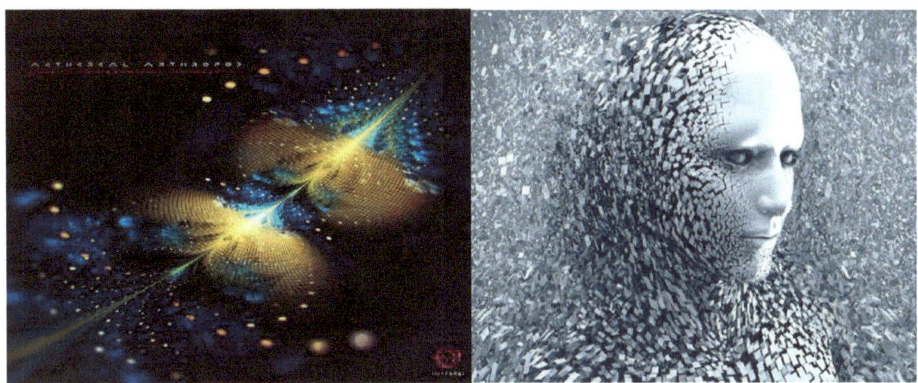

Existential Concept: Cloaking Behavior

Cloaking can be defined as the reflective mimicking of kinetic patterns, dynamics, textures, behaviors and cultures of energy facilitation; in order to exploitatively replicate the hue ambience of a given environment.

Counter-cloaking, in accordance with the aforementioned definition, is the ambient capacity to strategically induce other visual schema, also within a given environment.

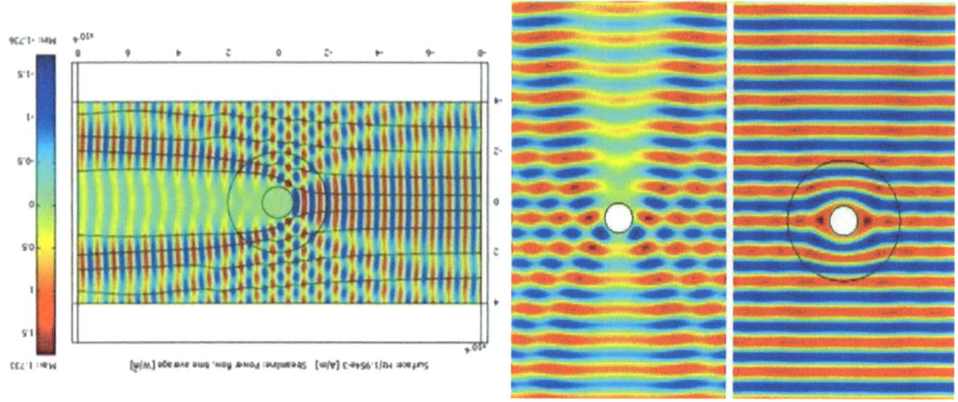

Existential Concept: Interstellar Teleportation

If we consider the Hermetic phrase: "As above, so below…," then the earthly pelagic gulf stream of hydro-current motion, must also have a reasonable interstellar consistency above; in order to facilitate the cosmic dynamics of interplanetary travel.

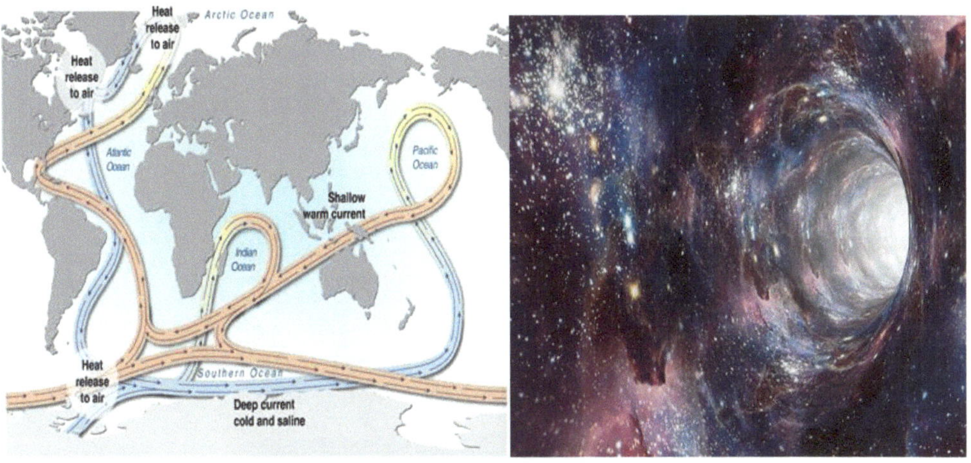

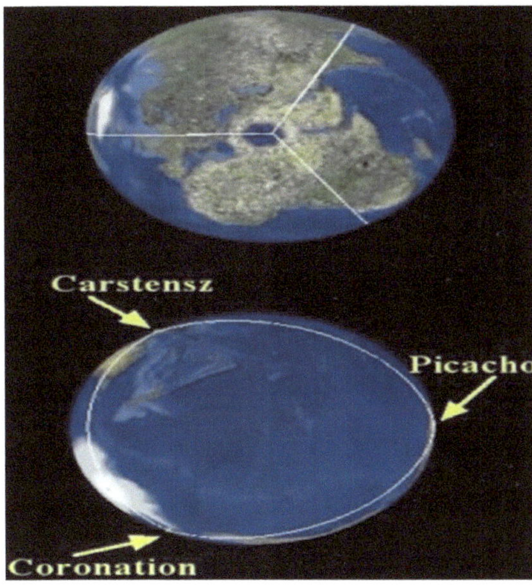

Existential Concept: Perception within Spatial Confinement

As it relates to the Singularity of the Oneness of existence, space can be defined as a confined or measured sense of orientation, within the established kinetic patterns or structures of dimensional consistency. Therefore, since it can be defined by measure, time can be a premier component to ubiquitously calculate its quantitative essence, as a relative point of reference.

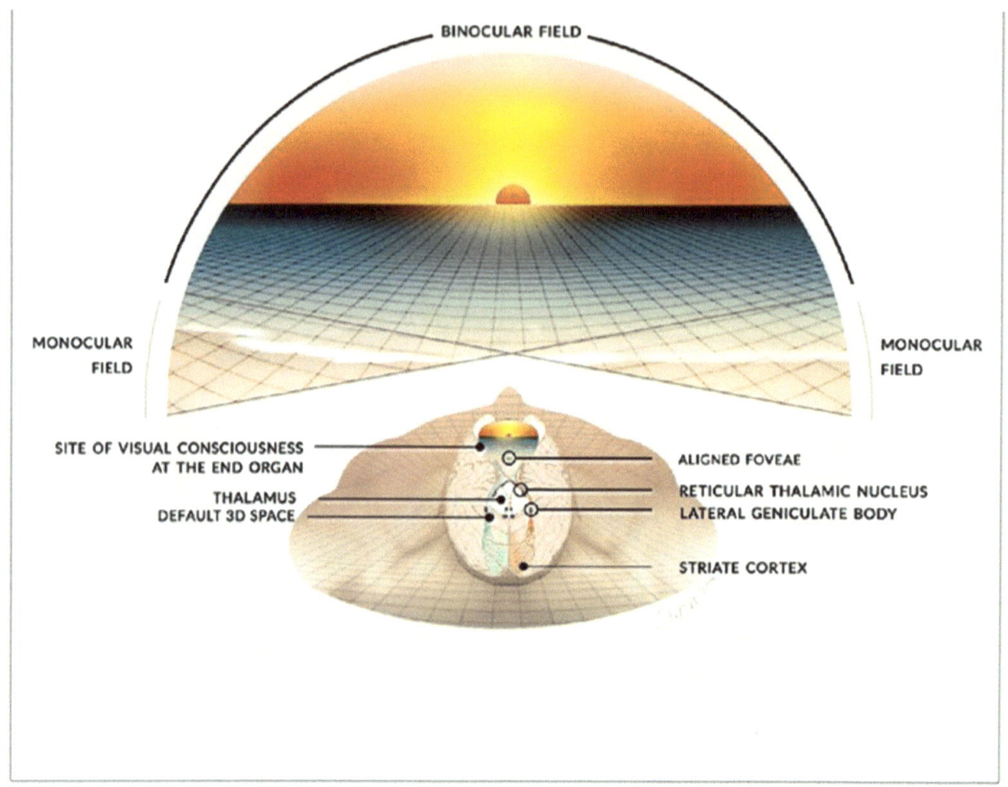

Existential Concept

Kinetic literacy can facilitate the fluidity of speed. Thus, speed can be defined or determined by how well a kinetic flow is perceived or interpreted within a given environment.

Dark Humor

Humorously, if asked, do I have some form of PTSD, I would have to say yes; for I am always prone to (P)otentially (T)hink (S)omewhat (D)ifferent.

Humorously, if anyone tells you that they are not afraid, they are either wise enough to acutely understand that it is utterly futile to be fearful; or they lack the intelligence to prudently understand, why it is not wise to recklessly tempt the no-nonsensical hand of fate.

Humorously, when civilization is fashioned by those of a low life outlook, who assume that they are in good standing, many may seek the euphoria of the high life; in order to cope with the overwhelming ineptitude of its complete state of dysfunction.

Amusingly, never fear those of a nefarious disposition, who profess to be fearless; until, they are faced with the daunting reality of having to change their ways.

Dark Theological Humor:

Magnate: "I am a wealthy venture capitalist, who is ready to fight the urge of indulging in the sin of graphic eroticism."

Chaplain: "Are you also willing to totally denounce yourself of the means which keep its industry lucrative?"

Magnate: "Maybe, what must I do?"

Chaplain: "First, study the motives that sponsor all fiscal entities; in order to understand that virtue, as it relates to capitalistic enterprises, typically makes hypocrites of the professed conscientious, who also yearn to have the intemperate affluence of proprietary gains.

Dark Theological Humor:

The Adversarial Mind: "Satan, do you fear death?"

Satan: "No, death is only a cyclic state; until, you awaken."

The Adversarial Mind: "So what gives you the incentive for being alive?"

Satan: "Hypocrisy, for it fosters the sense of doubt that I need, in order to authenticate that which can be considered the ultimate essence of freedom."

The Adversarial Mind: "Which is?"

Satan: "Pure unadulterated Truth."

Dark Humor: Nefarious Economics

Consumer: "Devil, I got a two dollar raise, now I am ready to buy that vase that you were selling for five dollars."

Devil: "Due to inflation, the price is now seven dollars."

Consumer: "That is a helluva price."

Devil: "No that is just the typical cost of capitalistic business."

Amusingly, any professed holy person, who claims to be chaste and pure, is either a latent philanderer, which has yet to experience puberty, or is under a psychosis of being afraid to show intimacy.

Humorously, if someone finds you, then ironically assume that you are lost and confused, be wary; for chances are they are truly lost and confused, but too proud to seek or ask for adequate direction.

Dark Humor:

Friend one: "What did you hear when you went to visit the wise man?"

Friend two: "I heard, hello."

Friend one: "Good, so he was welcoming,"

Friend two: "No, he had departed."

Friend one: "So how did you hear, hello?"

Friend Two: "It was me speaking, then hoping, but eventually realizing that I had taken him for granite."

Dark Humor

Financial Planner: "Wise man, it must be difficult to be wise and broke."

Wise man: "Not really, for the hubris and greed of your profession offers me the entrepreneurial means to be a crisis manager."

9 798815 119185